Scott Foresman - Addison Wesley

MATH

AF856337

Review From Last Year Masters

Grade 4

SF AW

Scott Foresman - Addison Wesley

Editorial Offices: Glenview, Illinois • New York, New York
Sales Offices: Reading, Massachusetts • Duluth, Georgia • Glenview, Illinois
Carrollton, Texas • Menlo Park, California

http://www.sf.aw.com

Overview

Review From Last Year Masters provides a review of key concepts from the previous year of Scott Foresman - Addison Wesley MATH. Each of the 20 masters begins with an instructional model that is followed by practice.

Review From Last Year Masters can be used to determine how well students have retained concepts from the previous year and to prepare students for the upcoming year.

The *Answers and Options for Further Review* section at the back of this book provides alternatives for students who need additional reteaching and practice. Materials from the previous year (Student Edition lessons, Reteaching Masters, and Practice Masters) are keyed to each master.

ISBN 0-201-48883-3

Printed in the United States of America

4 5 6 7 8 9 10 – CRK – 03 02 01

Contents

Name ____________________

Reading Graphs

The **bar graph** shows how many glasses of milk each student drinks per day.

How many glasses of milk does Ann drink each day?

Ann drinks 2 glasses of milk per day.

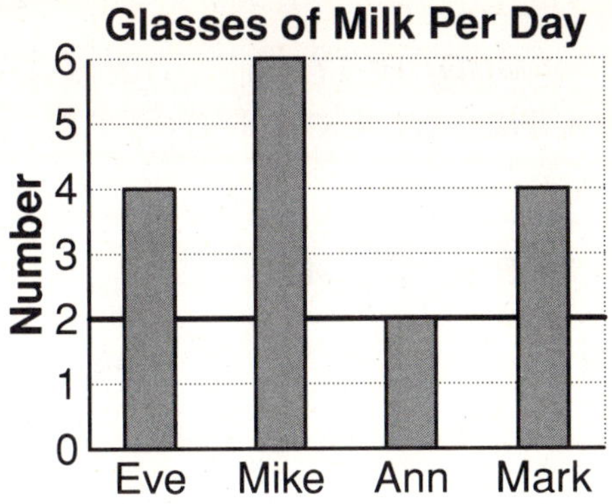

Use the graphs to answer each question.

1. How many books did Mark read? ________

2. Who read 6 books? ________

3. Who read the most books? ________

4. How many hours of sleep does a mole get each day? ________

5. Which creature sleeps about 4 hours each day? ________

6. Which creature sleeps the most each day? ________

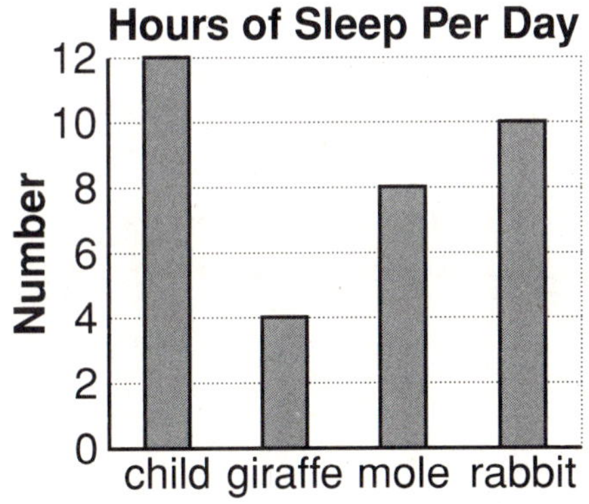

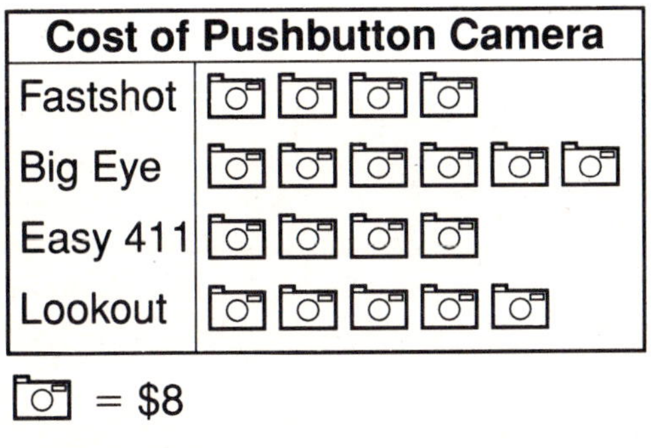

7. How much does the Lookout camera sell for? ________

8. Which camera sells for the same price as the Easy 411? ________

Name ________________________________

Review 2

Problem Solving: Analyze Word Problems

Add or subtract? Orlando is 14 years old. Asia is 11 years older than Orlando. How old is Asia?

The question says that Asia's age is 11 years more than Orlando's age. I will use **addition.**

Solution: **14 + 11 = 25. Asia is 25 years old.**

Write which operation you would use. Then solve.

1. A cheetah can run about 70 miles per hour. The fastest human can run about 27 miles per hour. How much faster is a cheetah than a human?

__________ ____________________

2. At the Premier Baseball Card Show, Jose bought one of each of the cards listed. How much did he spend?

Card Prices	
Barry Bonds	$24
Mark McGwire	15
Roger Clemens	18

__________ ____________________

3. At Operation Dogwash, students washed 38 dogs on Friday, 55 dogs on Saturday, and 19 dogs on Sunday. How many dogs were washed altogether?

__________ ____________________

4. Last year, the highest temperature in Lincoln City was 98°F. The lowest temperature was 16°F. What was the difference between the highest and lowest temperatures?

__________ ____________________

Name ____________________

Review 3

Place Value Through Hundred Thousands

Numbers can be written in three ways.

standard form	830,951
word form	**eight** hundred **thirty** thousand, **nine** hundred **fifty-one**
expanded form	800,000 + 30,000 + 900 + 50 + 1

Write the word form of each number.

1. 425,726 __________ hundred ____________________ thousand,

__________ hundred ____________________

2. 58,314 ____

____

3. 710, 504 ____

____

Write the standard form of each number.

4. two hundred ninety-nine thousand, six hundred thirty

____

5. sixty-eight thousand, two hundred fifty-four

____

Write the expanded form of each number.

6. 71,502 ____

7. 321,800 ____

Name ______________________________

Review 4

Adding Greater Numbers

Find 471 + 658.

Add the ones.	*Add the tens.* *Regroup.*	*Add the hundreds.*
47**1** + 65**8** **9**	1 4**7**1 + 6**5**8 **2**9	1 **4**71 + **6**58 **1,1**29

Find each sum

1. 6**3** + 9**4** **7**	**2.** 1 4**9** + 3**5** **4**	**3.** 55 + 88	**4.** 76 + 54
5. 133 + 274	**6.** 506 + 605	**7.** 462 + 87	**8.** 370 + 265
9. 187 + 783	**10.** 665 + 354	**11.** 900 + 447	**12.** 681 + 735
13. 40 624 + 315	**14.** 971 5 + 48	**15.** 338 93 + 776	**16.** 602 255 + 370

17. An iceberg stood 78 ft above water and 543 ft under water. How tall was the entire iceberg? ______________

Name ______________________________

Review 5

Subtracting Greater Numbers

Find 352 − 165.

Regroup from tens. Subtract the ones.	*Regroup from hundreds. Subtract the tens*	*Subtract the hundreds.*
4 12 3 5̸ 2̸ − 1 6 5 7	14 2 4̸ 12 3̸ 5̸ 2̸ − 1 6 5 8 7	14 2 4̸ 12 3̸ 5̸ 2̸ − 1 6 5 1 8 7

Find each difference.

1. 97 − 54 = 3 (ones shown)
2. 7 11 / 8̸1̸ − 35 = 6 (ones shown)
3. 47 − 19
4. 70 − 25
5. 248 − 137
6. 359 − 175
7. 506 − 70
8. 281 − 229
9. 648 − 375
10. 940 − 457
11. 400 − 91
12. 611 − 166
13. 803 − 74
14. 612 − 399
15. 1,478 − 549
16. 1,535 − 1,347

17. New York City's longest tunnels are the Brooklyn-Battery Tunnel (9,117 ft) and the Holland Tunnel (8,557 ft). Find the difference in the lengths of the tunnels. ______________

Name ______________________________

Review 6

Multiplication (0, 1, 2, 5, and 9 as factors)

0 Find **0** × 8. The product of 0 and any number is 0. 0 × 8 = 0	**1** Find **1** × 12. The product of 1 and any number is the number. 1 × 12 = 12	**2** Find 6 × **2**. Skip count by 2s. 0, 2, 4, 6, 8, 10, 12 6 × 2 = 12
5 Find 4 × **5**. Skip count by 5s. 0, 5, 10, 15, 20 4 × 5 = 20	**9** Find 4 × **9**. Use patterns. 4 × 9 = 36	

	tens	ones
1 × 9 =	**0**	**9**
2 × 9 =	**1**	**8**
3 × 9 =	**2**	**7**
4 × 9 =	**3**	**6**

Find each product.

1. 2 × 7 = ________ **2.** 0 × 5 = ________ **3.** 9 × 3 = ________

4. 5 × 6 = ________ **5.** 2 × 8 = ________ **6.** 1 × 8 = ________

7. 5 × 8 = ________ **8.** 0 × 6 = ________ **9.** 9 × 7 = ________

10. 2 × 4 = ________ **11.** 1 × 3 = ________ **12.** 9 × 8 = ________

13. 5 × 7 **14.** 6 × 2 **15.** 9 × 6 **16.** 7 × 0

17. 5 × 1 **18.** 5 × 5 **19.** 3 × 5 **20.** 1 × 9

21. Every birthday, Jed receives 9 silver dollars from his uncle. Jed is 9. How many silver dollars has he received from his uncle? ________

Name ______________________________

Review 7

Multiplication (3, 4, 6, and 7 as factors)

Solve simpler problems to multiply by 3, 4, 6, and 7.

3 Find **3** × 7. Find **2** × 7. **2** × 7 = 14 Find **1** × 7. + **1** × 7 = 7 Add. **3** × 7 = 21	**4** Find **4** × 9. Find **2** × 9. **2** × 9 = 18 Find **2** × 9. + **2** × 9 = 18 Add. **4** × 9 = 36
6 Find **6** × 8. Find **5** × 8. **5** × 8 = 40 Find **1** × 8. + **1** × 8 = 8 Add. **6** × 8 = 48	**7** Find **7** × 9. Find **5** × 9. **5** × 9 = 45 Find **2** × 9. + **2** × 9 = 18 Add. **7** × 9 = 63

Find each product.

1. 8 × 3 = ______ **2.** 6 × 2 = ______ **3.** 4 × 8 = ______

4. 7 × 7 = ______ **5.** 7 × 2 = ______ **6.** 3 × 3 = ______

7. 5 × 4 = ______ **8.** 6 × 6 = ______ **9.** 7 × 5 = ______

10. 6 × 7 = ______ **11.** 5 × 3 = ______ **12.** 4 × 4 = ______

13. 2 × 3 **14.** 6 × 5 **15.** 7 × 8 **16.** 1 × 6

17. 9 × 7 **18.** 4 × 2 **19.** 3 × 4 **20.** 9 × 6

21. There are 4 cups in a quart. How many cups are there in 6 quarts of milk? ______

Name ______________________________

Review 8

Division (by 2, 3, 4, and 5)

Use multiplication facts to find division facts.

2 Find 14 ÷ 2. What times 2 is 14? 2 × **7** = 14 14 ÷ 2 = **7**	**3** Find 24 ÷ 3. What times 3 is 24? 3 × **8** = 24 24 ÷ 3 = **8**
4 Find 12 ÷ 4. What times 4 is 12? 4 × **3** = 12 12 ÷ 4 = **3**	**5** Find 40 ÷ 5. What times 5 is 40? 5 × **8** = 40 40 ÷ 5 = **8**

Find each quotient.

1. 6 ÷ 2 = ______ **2.** 16 ÷ 4 = ______ **3.** 12 ÷ 2 = ______

4. 35 ÷ 5 = ______ **5.** 36 ÷ 4 = ______ **6.** 30 ÷ 5 = ______

7. 28 ÷ 4 = ______ **8.** 18 ÷ 2 = ______ **9.** 12 ÷ 3 = ______

10. $3\overline{)21}$ **11.** $5\overline{)40}$ **12.** $5\overline{)15}$

13. $2\overline{)14}$ **14.** $3\overline{)15}$ **15.** $3\overline{)27}$

16. $4\overline{)32}$ **17.** $2\overline{)10}$ **18.** $5\overline{)25}$

19. Margo spent $24 on tennis balls. Each can of balls cost $3. How many cans did she buy? ______

Name ___________________________________

Division (by 6, 7, 8, and 9)

Use multiplication facts to find division facts.

6 Find 24 ÷ 6. What times 6 is 24? $6 \times \mathbf{4} = 24$ $24 \div 6 = \mathbf{4}$	**7** Find 35 ÷ 7. What times 7 is 35? $7 \times \mathbf{5} = 35$ $35 \div 7 = \mathbf{5}$
8 Find 24 ÷ 8. What times 8 is 24? $8 \times \mathbf{3} = 24$ $24 \div 8 = \mathbf{3}$	**9** Find 63 ÷ 9. What times 9 is 63? $9 \times \mathbf{7} = 63$ $63 \div 9 = \mathbf{7}$

Find each quotient.

1. 24 ÷ 6 = ______ **2.** 48 ÷ 8 = ______ **3.** 48 ÷ 6 = ______

4. 36 ÷ 9 = ______ **5.** 16 ÷ 8 = ______ **6.** 72 ÷ 9 = ______

7. 64 ÷ 8 = ______ **8.** 12 ÷ 6 = ______ **9.** 56 ÷ 7 = ______

10. $7\overline{)28}$ **11.** $9\overline{)54}$ **12.** $9\overline{)27}$

13. $6\overline{)18}$ **14.** $7\overline{)49}$ **15.** $7\overline{)35}$

16. $8\overline{)40}$ **17.** $6\overline{)36}$ **18.** $9\overline{)81}$

19. Julian earned $7 for every box of greeting cards that he sold. He earned $63 total. How many boxes did he sell? ______

Name ______________________

Review 10

Problem Solving: Analyze Strategies

Bill's Wood Shop makes chairs. They can make 8 chairs in 1 hour, 16 chairs in 2 hours, and 24 chairs in 3 hours. A customer wants 56 chairs. How long will it take them to make the chairs?

Time (h)	1	2	3	4	5	6	7	8
Chairs	8	16	24	32	40	48	56	64

It will take them 7 hours to make 56 chairs.

Problem Solving Strategies

- Use Objects/Act It Out
- Draw a Picture
- Look For a Pattern
- Guess and Check
- Make an Organized List
- **Make a Table**
- Solve a Simpler Problem
- Work Backward

Solve each problem. You can use any method.

1. Jennie can make 3 pancakes with 1 egg, 6 pancakes with 2 eggs, and 9 pancakes with 3 eggs. How many pancakes can she make with 8 eggs? ______________

Eggs	1	2	3						
Pancakes	3	6	9						

2. If you stick 2 darts in the dart board, how many different scores can you get? ______________

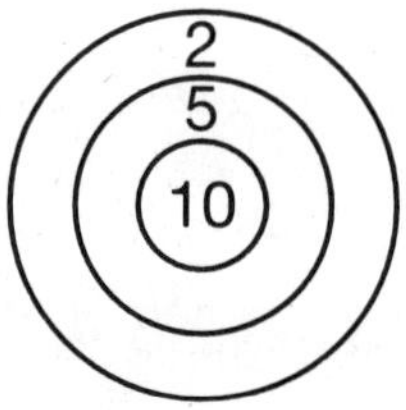

3. Charlie made 7 baskets in a basketball game. Some baskets were worth 2 points and some 3 points. He scored 16 points in all. How many 2-pointers and how many 3-pointers did he make? ______________

4. The house numbers on King Street follow the pattern shown. What will be the last number on the block? ______________

Name ______________________________

Review 11

Solids and Shapes

The solid is a rectangular prism.

The face of the solid is a rectangle.

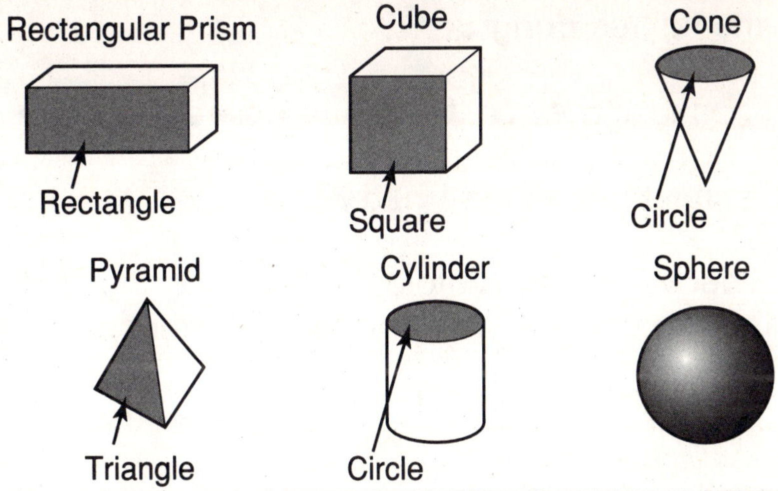

Name the shape of each face.

1. 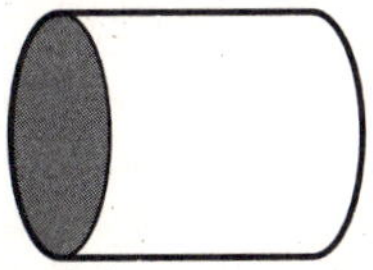

2. 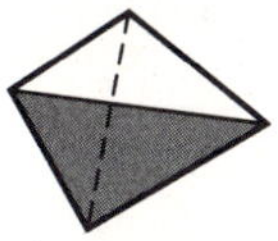

3. 

4. 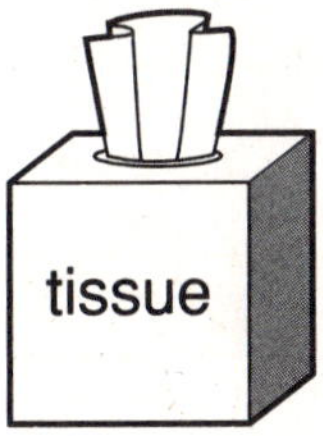

Name the solid.

5. 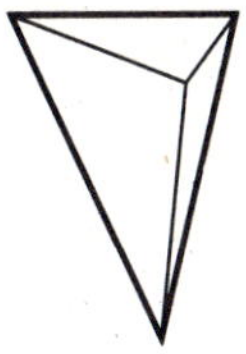

6. 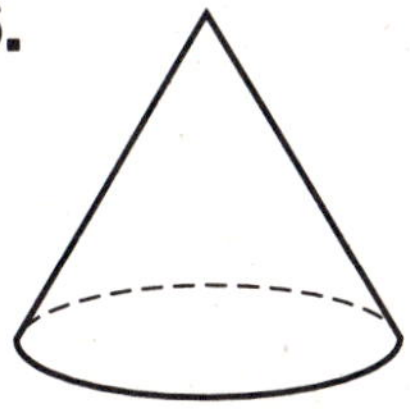

7. 

8. 

9. Name the solid which has these shapes as its faces.

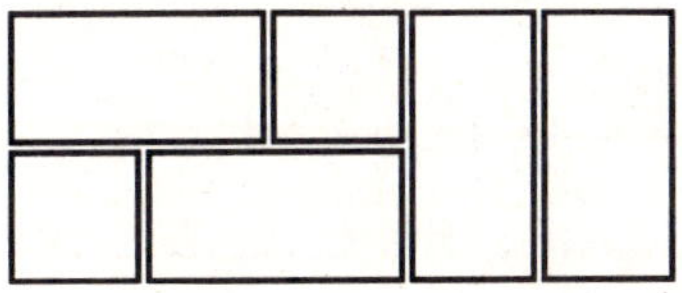

____

Name ______________________________

Review 12

Congruence and Symmetry

Are the figures **congruent**?

Yes, because they are the **same size and shape.**

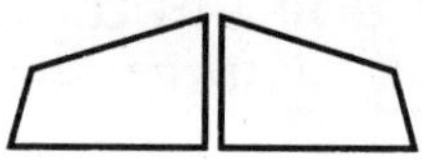

Is the line a **line of symmetry**?

Yes, because you can fold the figure so that **both parts match exactly.** The two parts are congruent.

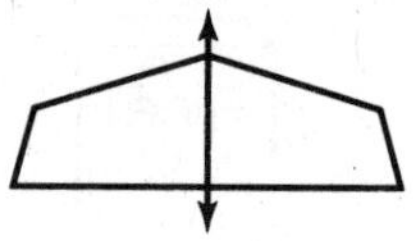

Are the figures congruent?

No. They are the same shape but not the same size.

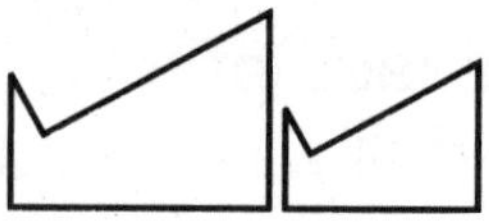

Tell if the figures are congruent. Write yes or no.

1. 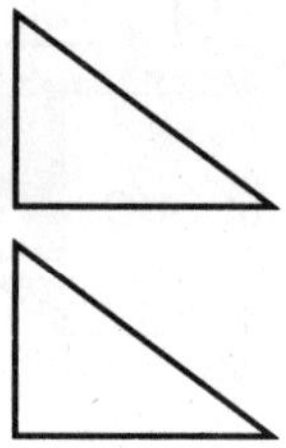 ______________

2. 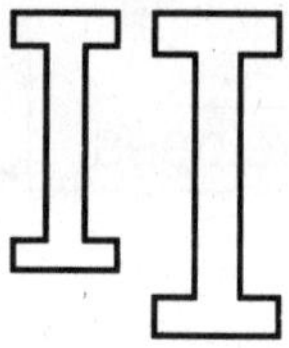 ______________

3. 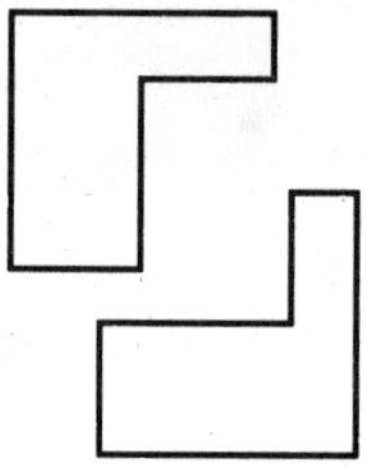 ______________

4. 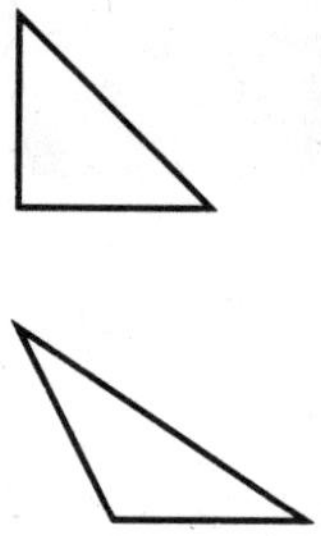 ______________

Tell if the line is a line of symmetry. Write yes or no.

5. 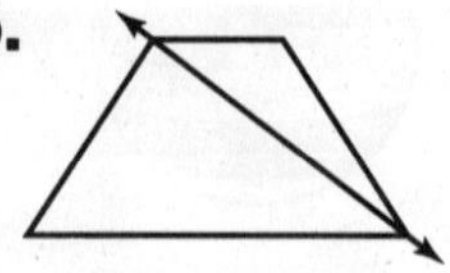 ______________

6. 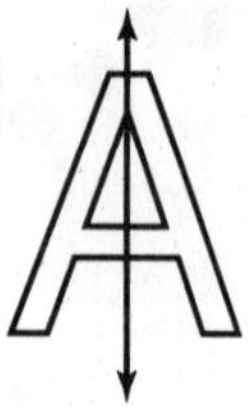 ______________

7. 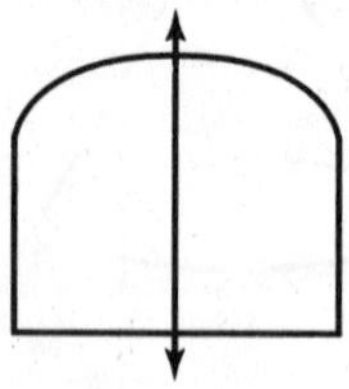 ______________

8. 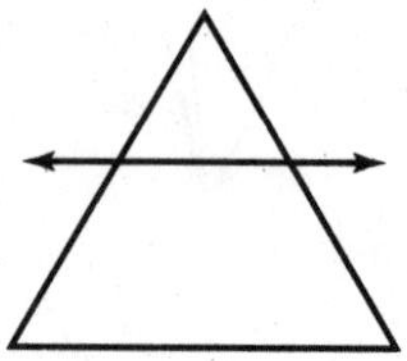 ______________

9. Use the line of symmetry to draw a congruent figure on the right side of the shape.

Name ______________________________

Exploring Perimeter and Area

Perimeter is the distance around a figure.

Perimeter = 3 + 2 + 3 + 2 = 10 units

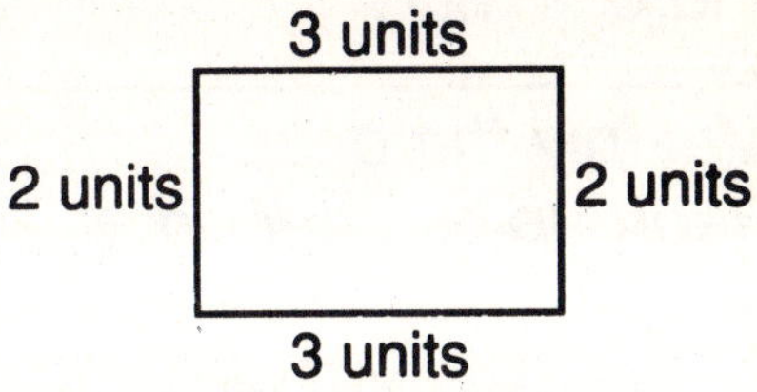

Area is the number of square units in a figure.

Area = 2 × 3 = 6 square units

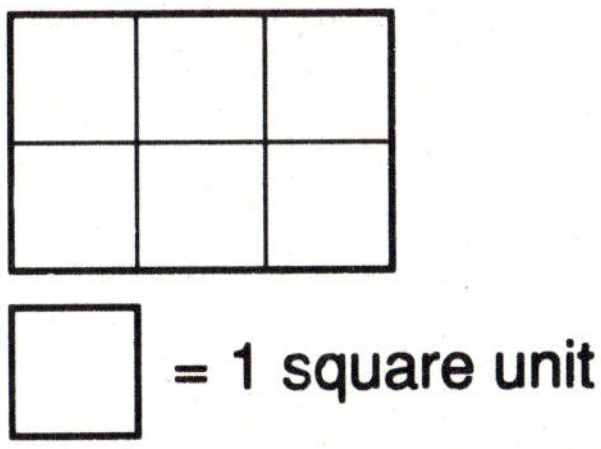

Complete.

1. What is the perimeter of the rectangle?

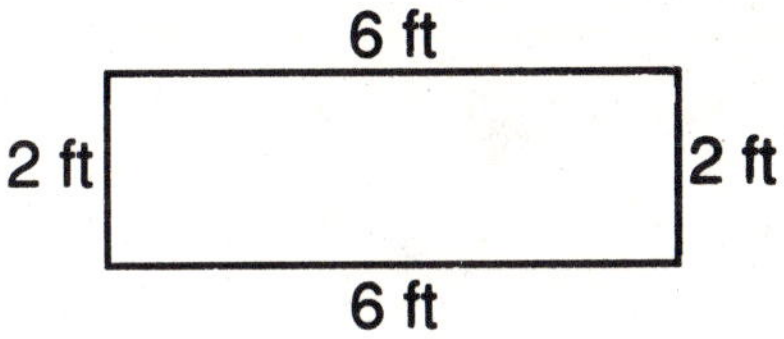

p = ______ + ______ + ______ + ______ = ___________

2. What is the area of the rectangle?

a = ______ × ______ = ___________

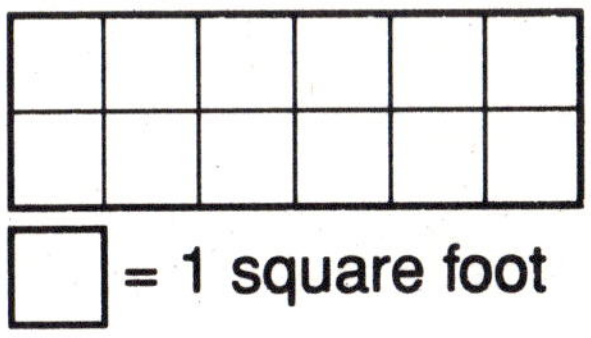

Find the perimeter and area of each figure.

3.

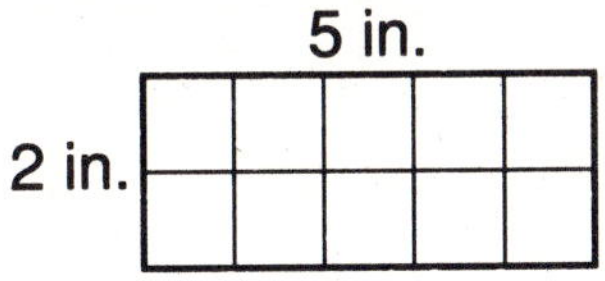

____

____

4.

____

____

5.

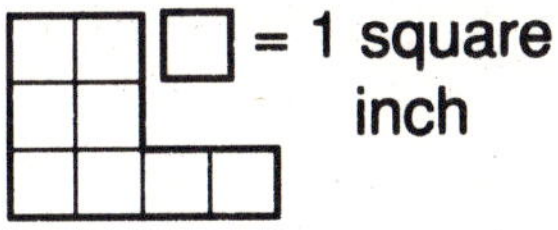

____

____

6.

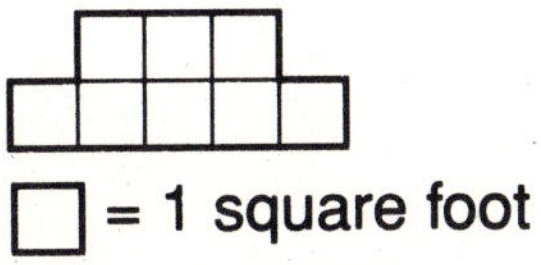

____

____

7.

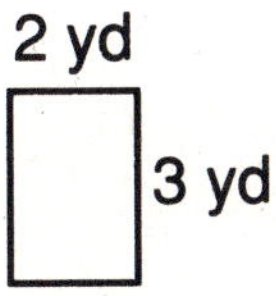

____

____

8.

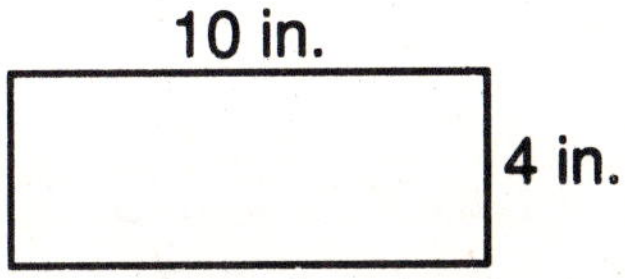

____

____

Name ______________________________

Review 14

Multiplying by 1-Digit Factors

Find 6 × 127.

Multiply the ones. Regroup as needed.	Multiply the tens. Add any extra tens. Regroup as needed.	Multiply the hundreds. Add any extra hundreds.
$\overset{4}{1}27$ ← 6 × 7 = [4]2 × 6 2 ← 6 × 7 = 4[2]	1 4 127 × 6 62	1 4 127 × 6 762

Find each product.

1. $\begin{array}{r} ^{1\,1}\\ 234 \\ \times \quad 3 \\ \hline \square 02 \end{array}$

2. $\begin{array}{r} ^{3\,1}\\ 162 \\ \times \quad 5 \\ \hline \square\square 0 \end{array}$

3. $\begin{array}{r} ^{2\,1}\\ 153 \\ \times \quad 5 \\ \hline \square\square\square \end{array}$

4. $\begin{array}{r} 491 \\ \times \quad 2 \\ \hline \end{array}$

5. $\begin{array}{r} 106 \\ \times \quad 7 \\ \hline \end{array}$

6. $\begin{array}{r} 204 \\ \times \quad 4 \\ \hline \end{array}$

7. $\begin{array}{r} 317 \\ \times \quad 3 \\ \hline \end{array}$

8. $\begin{array}{r} 249 \\ \times \quad 4 \\ \hline \end{array}$

9. $\begin{array}{r} 253 \\ \times \quad 5 \\ \hline \end{array}$

10. $\begin{array}{r} 361 \\ \times \quad 7 \\ \hline \end{array}$

11. $\begin{array}{r} 428 \\ \times \quad 6 \\ \hline \end{array}$

12. $\begin{array}{r} 380 \\ \times \quad 4 \\ \hline \end{array}$

13. The Acme Bus Company makes 8 buses every day. Last year the company was in operation 235 days. How many buses did it make? ____________

14. Fran bought 5 beach towels. Each towel cost $8.29. How much did she spend? ____________

Name ____________________

Review 15

Dividing by 1-Digit Divisors

Find 50 ÷ 6.

Estimate.	*Divide.*	*Subtract.*	*Compare. Write the remainder.*
$6\overline{)50}$ = 8 **Think:** 50 is close to 48. 48 ÷ 6 = 8	$6\overline{)50}$ = 8 48 ← 8 × 6	$6\overline{)50}$ = 8 48 2 ← remainder	$6\overline{)50}$ = 8 R2 48 2 ← 2 < 6

Find each quotient.

1. $7\overline{)2\ \ 4}$ 3 R ☐; 2 1; ☐

2. $4\overline{)3\ \ 3}$ R ☐; ☐ ☐; ☐

3. $5\overline{)6\ \ 7}$ R ☐

4. $3\overline{)2\ \ 5}$ R ☐

5. $4\overline{)5\ 7}$

6. $8\overline{)5\ 8}$

7. $9\overline{)4\ 9}$

8. $7\overline{)3\ 6}$

9. $6\overline{)1\ 4}$

10. $8\overline{)7\ 5}$

11. $6\overline{)6\ 2}$

12. $6\overline{)3\ 4}$

13. Mrs. Delia had 49 apples. She used 6 apples to make each jar of applesauce. How many apples were left over? ____________

14. John had 51 basketball cards. He sorted them into 6 equal piles. How many did he have left over? ____________

Name ______________________________

Review 16

Fractions

Write the equivalent fractions shown.

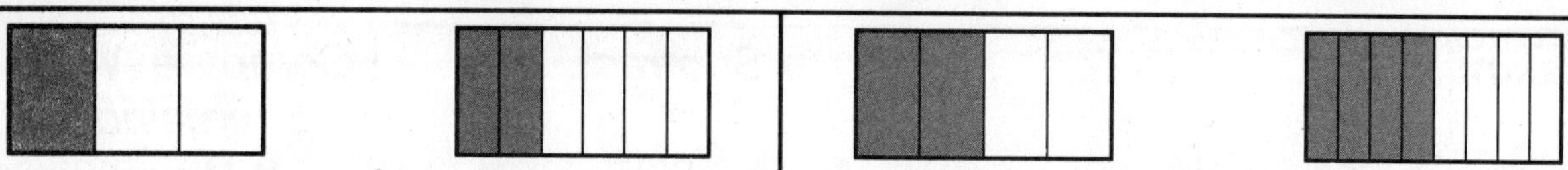

1 part shaded 3 parts in all		2 parts shaded 6 parts in all	2 parts shaded 4 parts in all		4 parts shaded 8 parts in all
$\frac{1}{3}$	=	$\frac{2}{6}$	$\frac{2}{4}$	=	$\frac{4}{8}$

Write the equivalent fractions shown.

1.

☐ parts shaded ☐ parts shaded
☐ parts in all ☐ parts in all

$\frac{\square}{\square} = \frac{\square}{\square}$

2.

☐ parts shaded ☐ parts shaded
☐ parts in all ☐ parts in all

$\frac{\square}{\square} = \frac{\square}{\square}$

3.

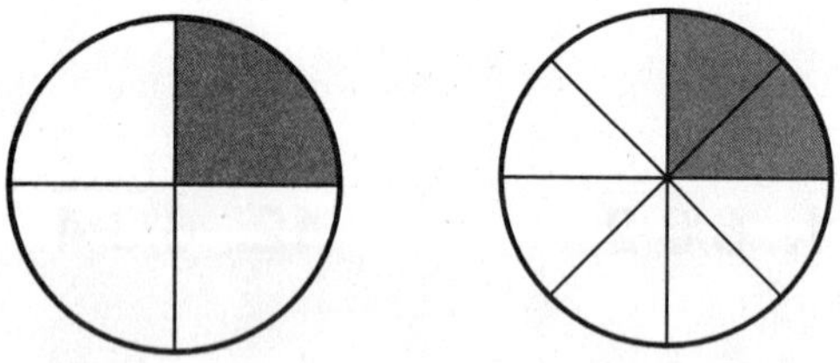

______ = ______

4.

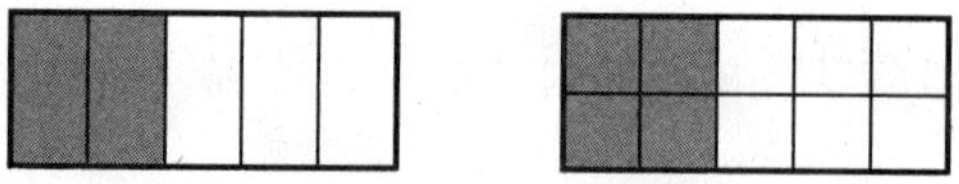

______ = ______

5.

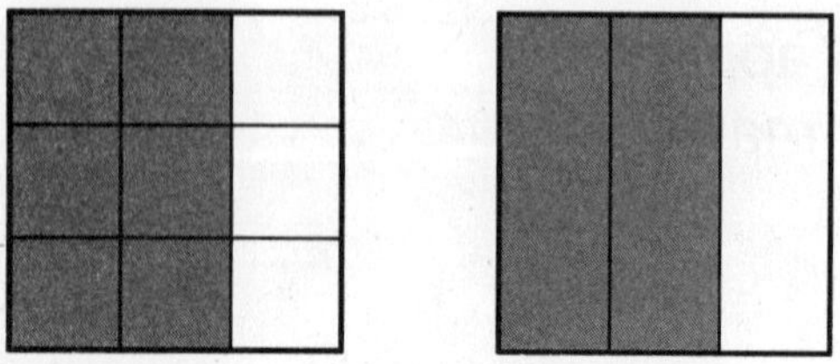

______ = ______

6.

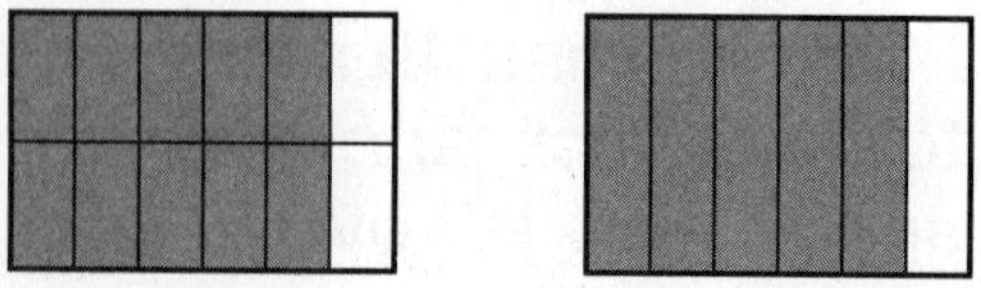

______ = ______

Name ____________________

Comparing Fractions

Compare $\frac{1}{4}$ and $\frac{2}{5}$.

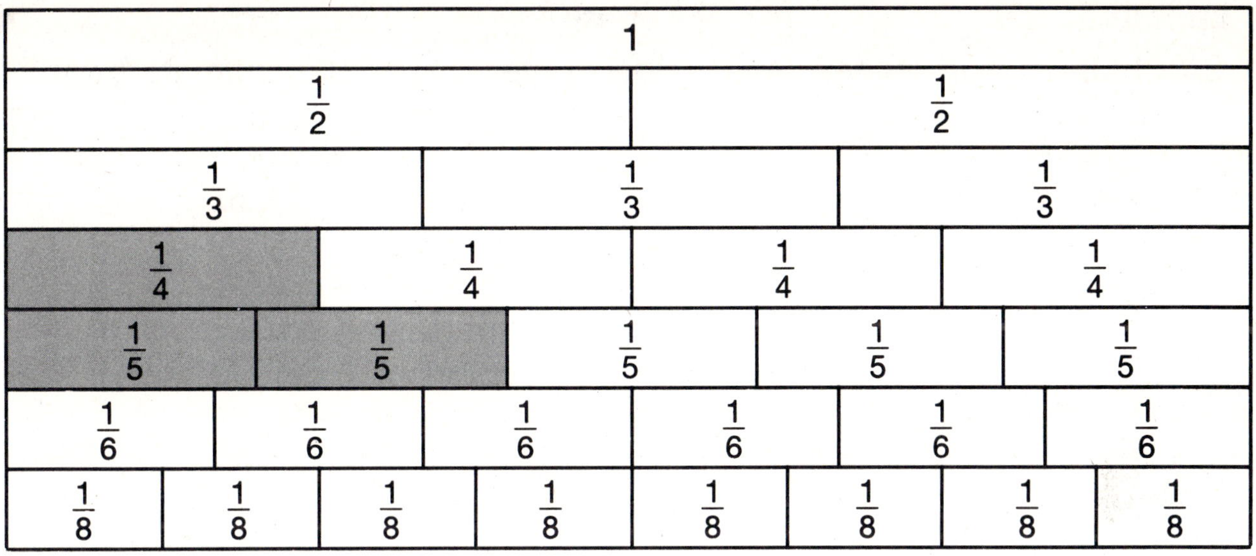

One $\frac{1}{4}$ rectangle is shorter than two $\frac{1}{5}$ rectangles. So, $\frac{1}{4} < \frac{2}{5}$.

Compare. Write <, >, or = in the square.

1. $\frac{1}{4}$ ☐ $\frac{1}{5}$

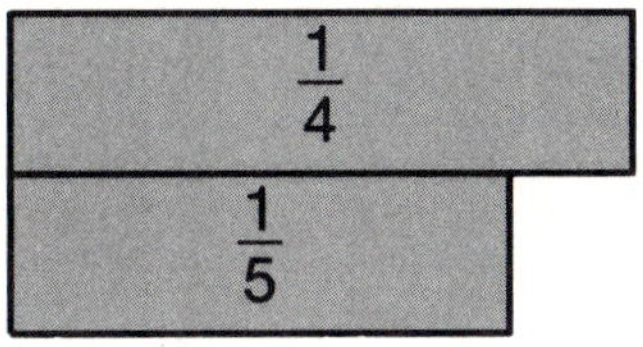

2. $\frac{1}{5}$ ☐ $\frac{2}{8}$

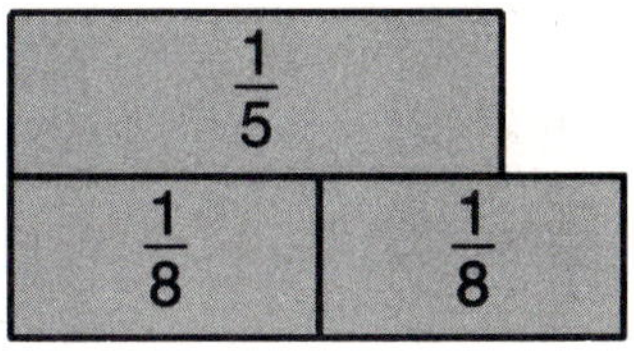

3. $\frac{1}{3}$ ☐ $\frac{1}{4}$

4. $\frac{1}{8}$ ☐ $\frac{1}{6}$

5. $\frac{1}{3}$ ☐ $\frac{2}{5}$

6. $\frac{1}{5}$ ☐ $\frac{2}{6}$

7. $\frac{1}{2}$ ☐ $\frac{2}{3}$

8. $\frac{2}{6}$ ☐ $\frac{1}{4}$

9. $\frac{1}{3}$ ☐ $\frac{2}{6}$

10. $\frac{2}{4}$ ☐ $\frac{1}{3}$

11. $\frac{1}{2}$ ☐ $\frac{3}{6}$

12. Dawn painted $\frac{3}{8}$ of the fence. Jed painted $\frac{1}{3}$ of the fence. Who painted more? ____________

Name ____________________

Review 18

Measurement

Customary Units of Measure		Metric Units of Measure	
Length		**Length**	
1 foot (ft)	= 12 inches (in.)	1 decimeter (dm)	= 10 centimeters (cm)
1 yard (yd)	= 36 in.	1 meter (m)	= 100 cm
	= 3 ft		= 10 dm
1 mile (mi)	= 5,280 ft	1 kilometer (km)	= 1,000 m
	= 1,760 yd		
Weight		**Mass**	
1 pound (lb)	= 16 ounces (oz)	1 kilogram (kg)	= 1,000 grams (g)

Compare. Write $<$, $>$, or $=$.

2 yd ______ 9 ft

Think: 1 yard = 3 feet.

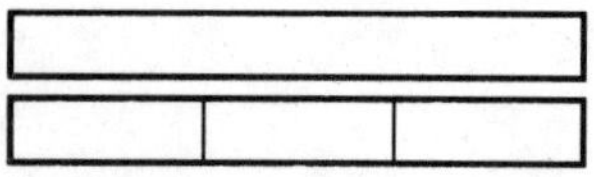

2 yd = 2 × 3 = 6 ft

2 yd $<$ 9 ft

1. 3 m ______ 200 dm

Think: 1 meter = ______ decimeters.

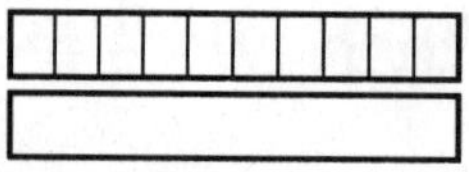

3 m = ______ × ______ = ______ dm

3 m ______ 200 dm

Compare. Write $<$, $>$, or $=$.

2. 2 ft ______ 24 in. **3.** 1 dm ______ 15 cm **4.** 5,280 ft ______ 1 mi

5. 2,000 g ______ 1 kg **6.** 1 lb ______ 16 oz **7.** 3 yd ______ 10 ft

Choose an estimate for each. Write the letter.

8. length of an ant ______ **a.** 1 g

9. weight of a dog ______ **b.** 200 ft

10. length of an airplane ______ **c.** 20 lb

11. weight of a paper clip ______ **d.** 1 cm

12. A bullfrog jumped 1 foot, 2 inches.
How many total inches did the bullfrog jump? ______

Name ______________________________

Review 19

Decimals

Write the fraction, decimal, and word name for each shaded part.

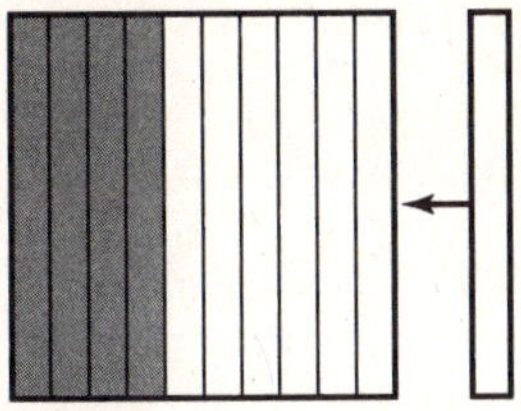

one tenth

0.1

$\frac{1}{10}$

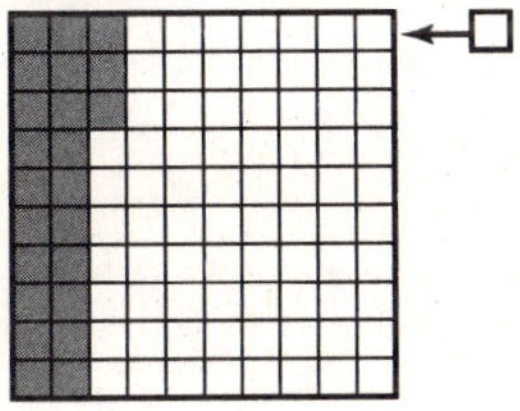

one hundredth

0.01

$\frac{1}{100}$

Fraction	Decimal	Word
$\frac{4}{10}$	0.4	four tenths
$\frac{23}{100}$	0.23	twenty-three hundredths

Write the fraction, decimal, and word name for each shaded part.

		Fraction	Decimal	Word
1.	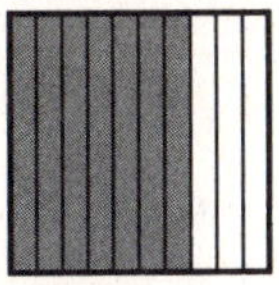	________	________	____________________
2.	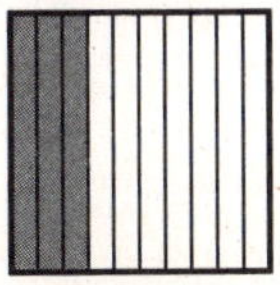	________	________	____________________
3.		________	________	____________________
4.	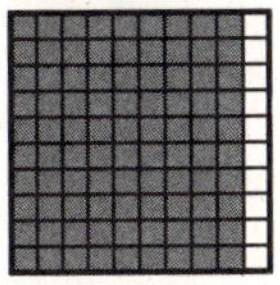	________	________	____________________

5. Shade the grid so that it shows twenty hundredths.

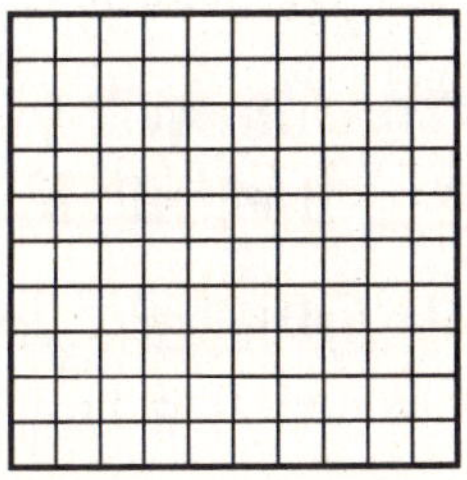

Name ______________________________

Probability

What is the probability of spinning an A?

1 ← number of As
4 ← number of equal sections

The probability of spinning an A is $\frac{1}{4}$.

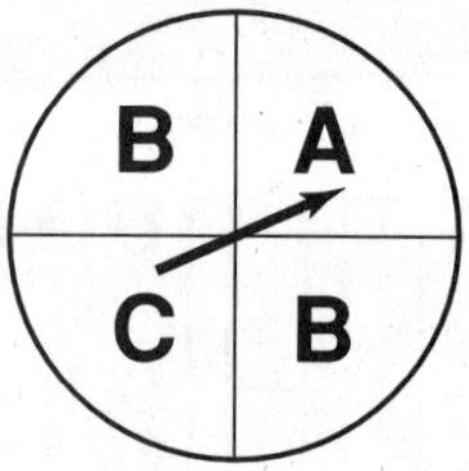

Find each probability.

1. What is the probability of spinning a C?

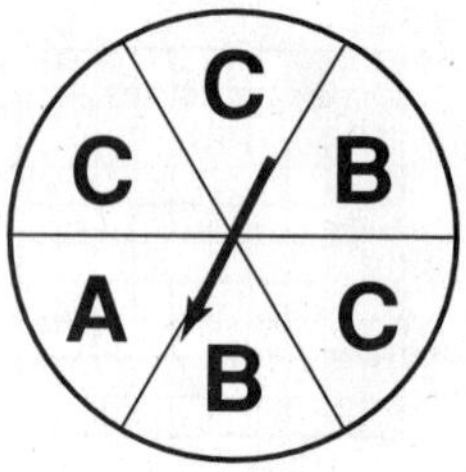

number of C → ☐

number of equal sections → ☐

The probability of spinning a C is __________.

2. What is the probability of spinning a B or C?

number of Bs and Cs → ☐

number of equal sections → ☐

The probability of spinning a B or C is __________.

What is the probability that you will pick each type of marble?

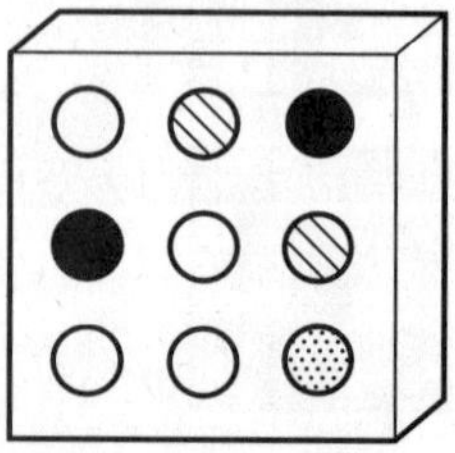

3. A black marble __________

4. A white marble __________

5. A striped marble __________

6. A spotted marble __________

7. Which marble will you most likely pick from the bag? ____________________

8. Which marble are you least likely to pick from the bag? ____________________

Answers and Options for Further Review

REVIEW 1
If students need more help on reading and making graphs, use Lessons 1-1 through 1-3 and 1-7 through 1-9 from the 3rd-grade book, along with the corresponding Practice and Reteaching Masters.

1. 4 books	**2.** Eve	**3.** Mike
4. 8 hours	**5.** giraffe	**6.** child
7. $40	**8.** Fastshot	

REVIEW 2
If students need more help on analyzing word problems, use Lesson 1-5 from the 3rd-grade book, along with the corresponding Practice and Reteaching Masters.

1. subtraction; 43 miles per hour
2. addition; $57
3. addition; 112 dogs
4. subtraction; 82°F

REVIEW 3
If students need more help on place value through hundred thousands, use Lesson 2-4 from the 3rd-grade book, along with the corresponding Practice and Reteaching Masters.

1. four hundred twenty-five thousand, seven hundred twenty-six
2. fifty-eight thousand, three hundred fourteen
3. seven hundred ten thousand, five hundred four
4. 299,630
5. 68,254
6. 70,000 + 1,000 + 500 + 2
7. 300,000 + 20,000 + 1,000 + 800

REVIEW 4
If students need more help on adding greater numbers, use Lessons 3-5 through 3-8 from the 3rd-grade book, along with the corresponding Practice and Reteaching Masters.

1. 157	**2.** 84	**3.** 143
4. 130	**5.** 407	**6.** 1,111
7. 549	**8.** 635	**9.** 970
10. 1,019	**11.** 1,347	**12.** 1,416
13. 979	**14.** 1,024	**15.** 1,207
16. 1,227	**17.** 621 ft	

REVIEW 5
If students need more help on subtracting greater numbers, use Lessons 4-6 through 4-12 from the 3rd-grade book, along with the corresponding Practice and Reteaching Masters.

1. 43	**2.** 46	**3.** 28
4. 45	**5.** 111	**6.** 184
7. 436	**8.** 52	**9.** 273
10. 483	**11.** 309	**12.** 445
13. 729	**14.** 213	**15.** 929
16. 188	**17.** 560 ft	

REVIEW 6
If students need more help on multiplication with 0, 1, 2, 5, and 9 as factors, use Lessons 5-4 through 5-8 from the 3rd-grade book, along with the corresponding Practice and Reteaching Masters.

1. 14	**2.** 0	**3.** 27
4. 30	**5.** 16	**6.** 8
7. 40	**8.** 0	**9.** 63
10. 8	**11.** 3	**12.** 72
13. 35	**14.** 12	**15.** 54
16. 0	**17.** 5	**18.** 25
19. 15	**20.** 9	

21. 81 silver dollars

REVIEW 7

If students need more help on multiplication with 3, 4, 6, 7, and 8 as factors, use Lessons 6-1 through 6-4 from the 3rd-grade book, along with the corresponding Practice and Reteaching Masters.

1. 24	**2.** 12	**3.** 32
4. 49	**5.** 14	**6.** 9
7. 20	**8.** 36	**9.** 35
10. 42	**11.** 15	**12.** 16
13. 6	**14.** 30	**15.** 56
16. 6	**17.** 63	**18.** 8
19. 12	**20.** 54	
21. 24 cups		

REVIEW 8

If students need more help on division by 2, 3, 4, and 5, use Lessons 7-5 through 7-7 from the 3rd-grade book, along with the corresponding Practice and Reteaching Masters.

1. 3	**2.** 4	**3.** 6
4. 7	**5.** 9	**6.** 6
7. 7	**8.** 9	**9.** 4
10. 7	**11.** 8	**12.** 3
13. 7	**14.** 5	**15.** 9
16. 8	**17.** 5	**18.** 5
19. 8 cans		

REVIEW 9

If students need more help on division by 6, 7, 8, and 9, use Lessons 7-10 and 7-11 from the 3rd-grade book, along with the corresponding Practice and Reteaching Masters.

1. 4	**2.** 6	**3.** 8
4. 4	**5.** 2	**6.** 8
7. 8	**8.** 2	**9.** 8
10. 4	**11.** 6	**12.** 3
13. 3	**14.** 7	**15.** 5
16. 5	**17.** 6	**18.** 9
19. 9 boxes		

REVIEW 10

If students need more help on the strategy Making a Table, use Lesson 9-10 from the 3rd-grade book, along with the corresponding Practice and Reteaching Masters.

1. 24 pancakes
2. 6 scores (4, 7, 10, 12, 15, 20)
3. 5 2-pointers and 2 3-pointers
4. 452

REVIEW 11

If students need more help on solids and shapes, use Lessons 8-1 and 8-2 from the 3rd-grade book, along with the corresponding Practice and Reteaching Masters.

1. circle	**2.** triangle
3. rectangle	**4.** square
5. pyramid	**6.** cone
7. cube	**8.** sphere
9. rectangular prism	

REVIEW 12

If students need more help on congruence and symmetry, use Lesson 8-6 from the 3rd-grade book, along with the corresponding Practice and Reteaching Masters.

1. yes	**2.** no	**3.** yes	**4.** no
5. no	**6.** yes	**7.** yes	**8.** no

9.

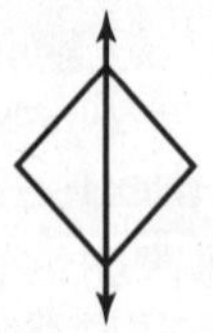

REVIEW 13

If students need more help on perimeter and area, use Lessons 8-8 and 8-9 from the 3rd-grade book, along with the corresponding Practice and Reteaching Masters.

1. 16 feet
2. 12 square feet
3. p = 14 inches; a = 10 square inches
4. p = 16 feet; a = 7 square feet
5. p = 14 inches; a = 8 square inches
6. p = 14 feet; a = 8 square feet
7. p = 10 yards; a = 6 square yards
8. p = 28 inches; a = 40 square inches

REVIEW 14

If students need more help on multiplying by 1-digit factors, use Lessons 9-5 through 9-7 from the 3rd-grade book, along with the corresponding Practice and Reteaching Masters.

1. 702	**2.** 810	**3.** 765
4. 982	**5.** 742	**6.** 816
7. 951	**8.** 996	**9.** 1,265
10. 2,527	**11.** 2,568	**12.** 1,520
13. 1,880 buses		**14.** $41.45

REVIEW 15

If students need more help on dividing by 1-digit divisors, use Lessons 9-11 through 9-14 from the 3rd-grade book, along with the corresponding Practice and Reteaching Masters.

1. 3 R3	**2.** 8 R1	**3.** 13 R2
4. 8 R1	**5.** 14 R1	**6.** 7 R2
7. 5 R4	**8.** 5 R1	**9.** 2 R2
10. 9 R3	**11.** 10 R2	**12.** 5 R4
13. 1 apple	**14.** 3 cards	

REVIEW 16

If students need more help on fractions, use Lesson 10-2 from the 3rd-grade book, along with the corresponding Practice and Reteaching Masters.

1. $\frac{2}{3} = \frac{4}{6}$	**2.** $\frac{3}{4} = \frac{6}{8}$	**3.** $\frac{1}{4} = \frac{2}{8}$
4. $\frac{2}{5} = \frac{4}{10}$	**5.** $\frac{6}{9} = \frac{2}{3}$	**6.** $\frac{10}{12} = \frac{5}{6}$

REVIEW 17

If students need more help on comparing fractions, use Lesson 10-4 from the 3rd-grade book, along with the corresponding Practice and Reteaching Masters.

1. >	**2.** <	
3. >	**4.** <	**5.** <
6. <	**7.** <	**8.** >
9. =	**10.** >	**11.** =
12. Dawn		

REVIEW 18

If students need more help on measurement, use Lessons 10-11 through 10-14 and 11-6 through 11-7 from the 3rd-grade book, along with the corresponding Practice and Reteaching Masters.

1. <	**2.** =	**3.** <
4. =	**5.** >	**6.** =
7. <	**8.** d	**9.** c
10. b	**11.** a	
12. 14 inches		

REVIEW 19

If students need more help on decimals, use Lessons 11-1 and 11-2 from the 3rd-grade book, along with the corresponding Practice and Reteaching Masters.

1. $\frac{7}{10}$, 0.7, seven-tenths
2. $\frac{3}{10}$, 0.3, three-tenths
3. $\frac{67}{100}$, 0.67, sixty-seven hundredths
4. $\frac{90}{100}$ (or $\frac{9}{10}$), 0.90 (or 0.9), ninety-hundredths (or nine-tenths)
5. 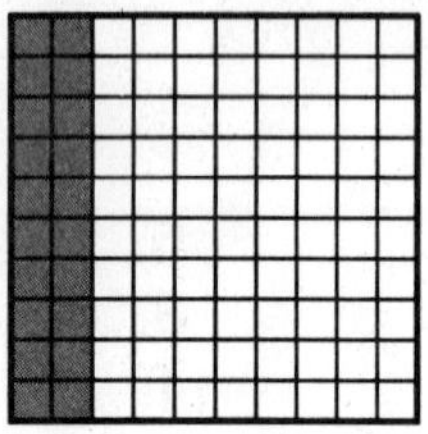

REVIEW 20

If students need more help on probability, use Lesson 12-9 from the 3rd-grade book, along with the corresponding Practice and Reteaching Masters.

1. $\frac{3}{6}$ (or $\frac{1}{2}$)
2. $\frac{5}{8}$
3. $\frac{2}{9}$
4. $\frac{4}{9}$
5. $\frac{2}{9}$
6. $\frac{1}{9}$
7. white
8. spotted